discovermore

Marine Wildlife

Mollusks

Kaitlyn Salvatore

Britannica
Educational Publishing

IN ASSOCIATION WITH

THE ROSEN PUBLISHING GROUP

Published in 2025 by Britannica Educational Publishing (a trademark of Encyclopædia Britannica, Inc.) in association with The Rosen Publishing Group, Inc.
2544 Clinton Street, Buffalo, NY 14224

Distributed exclusively by Rosen Publishing.
To see additional Britannica Educational Publishing titles, go to rosenpublishing.com.

Editor: Brianna Propis
Book Design: Michael Flynn

Photo Credits: Cover Shaf/Shutterstock.com; (series background) Dai Yim/Shutterstock.com; p. 4 Al Carrera/Shutterstock.com; p. 5 Damsea/Shutterstock.com; p. 6 Svitlyk/Shutterstock.com; p. 7 Victor1153/Shutterstock.com; p. 8 Alexey Masliy/Shutterstock.com; p. 9 KittyVector/Shutterstock.com; p. 11 (top) Arunee Rodloy/Shutterstock.com; p. 11 (bottom) Nina Milton/Shutterstock.com; p. 12 allstars/Shutterstock.com; p. 13 (left) Japan's Fireworks/Shutterstock.com; p. 13 (right) Shpatak/Shutterstock.com; p. 14 Divelvanov/Shutterstock.com; p. 15 LuckyStep/Shutterstock.com; p. 16 AndreyShutov/Shutterstock.com; p. 17 elena moiseeva/Shutterstock.com; p. 19 (top) Tomasz Klejdysz/Shutterstock.com; p. 19 (bottom) George P Gross/Shutterstock.com; p. 21 (top) kkshxt/Shutterstock.com; p. 21 (bottom) SNC Art and More/Shutterstock.com; p. 22 ATTOMY/Shutterstock.com; p. 23 Richard Whitcombe/Shutterstock.com; p. 24 Brent Hofacker/Shutterstock.com; p. 25 Merrillie Redden/Shutterstock.com; p. 26 Sameer Neamah Mahdi/Shutterstock.com; p. 27 pang_oasis/Shutterstock.com; p. 28 Kichigin/Shutterstock.com; p. 29 ohrim/Shutterstock.com.

Library of Congress Cataloging-in-Publication Data

Names: Salvatore, Kaitlyn, author.
Title: Mollusks / Kaitlyn Salvatore.
Description: [Buffalo, New York] : Britannica Educational Publishing, [2025] | Series: Discover more marine wildlife | Includes bibliographical references and index.
Identifiers: LCCN 2024033046 | ISBN 9781641903578 (library binding) | ISBN 9781641903561 (paperback) | ISBN 9781641903585 (ebook)
Subjects: LCSH: Mollusks--Juvenile literature.
Classification: LCC QL405.2 .S247 2025 | DDC 594--dc23/eng/20240731
LC record available at https://lccn.loc.gov/2024033046

Manufactured in the United States of America

CPSIA Compliance Information: Batch #CWBRIT25. For further information contact Rosen Publishing at 1-800-237-9932.

Contents

The Mollusk Squad

Earth is home to many different creatures. From humpback whales to bumblebees to bears, **habitats** around the world are full of amazing animals that are interesting and diverse—including mollusks. Mollusks are the second-largest phylum, or group, of animals, with about 100,000 species, or kinds. Each species of mollusk has similarities and differences to other mollusks. Most mollusks only live in water, but some live on the land. A lot of mollusks have a shell on their back, but some do not. Snails and squid are both mollusks, but they are very different from each other!

The cuttlefish is one of millions of mollusks that call our planet home.

Mollusks live in nearly all freshwater and marine environments. Some live in shallow waters, and others live in the deepest parts of the ocean.

Millions of mollusks live on Earth in different environments all around the world. They have special characteristics that help them survive and set them apart from other animals—and each other. The world of mollusks is as fascinating as it is huge!

WORD WISE

A HABITAT IS THE NATURAL HOME OR ENVIRONMENT OF AN ANIMAL, PLANT, OR OTHER LIVING THING.

What Is a Mollusk?

A mollusk is a type of animal with a soft, unsegmented (not in parts) body. Most mollusks have a hard shell that protects their body, but not all do. There are more than 100,000 species of mollusks in the world. Scientists think they account for 23 percent of all known sea animals.

The Carpathian blue slug is a large land slug. It creeps around on grass, moss, and piles of leaves.

About 95 percent of all animals on Earth are invertebrates!

With so many mollusk species, it's no surprise how different they are. But they do have some traits, or things, in common. Mollusks are invertebrates, which means they do not have a backbone, and their bodies are not divided into clear sections. Generally, mollusks have three body regions—a head, a visceral mass, and a footlike structure. The head contains the brain. The visceral mass contains the organs. It is surrounded by a protective layer called the mantle. In some mollusks, the mantle forms the animal's shell. Its "foot" helps it to dig, grasp, and creep around.

Consider This

Why would a hard shell be helpful to a soft-bodied slug?

There are seven classes, or groups, of mollusks grouped by their similar traits. Even mollusks in the same class are varied. They can be different sizes and colors, behave differently, and live in different habitats.

The bivalves include oysters, clams, and mussels. These mollusks have a shell with two halves (also called valves) and a soft body. Snails, slugs, and conches are known as gastropods. Most gastropods have a shell, although some do not. They also have a muscular foot, and many have a ring of teeth called a radula. About 65 percent of mollusks belong to this class.

Octopuses, squid, cuttlefish, and nautiluses are in a group called cephalopods. Cephalopods swim and generally do not have shells. They have the most developed brain and eyes of all mollusks. This class of mollusks contains the biggest members.

Clams are bivalve mollusks. Their two shells help protect them from being easily eaten by predators such as octopuses.

Mollusca

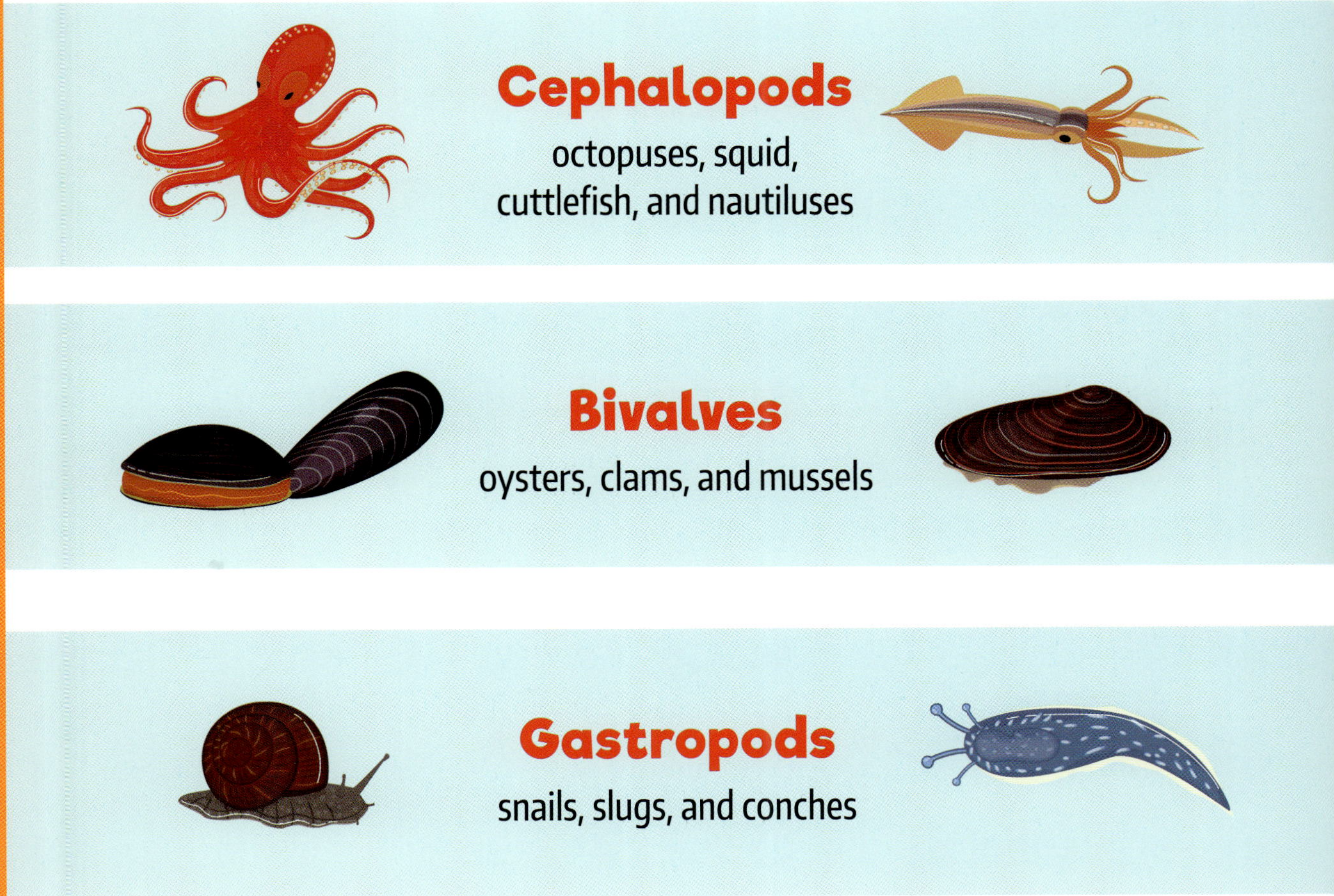

This illustration shows some mollusk classes and the animals within them.

Consider This

Why might the largest mollusks in the cephalopod group not have shells? Would they have less of a need for them?

Bivalve Bodies and Behaviors

Bivalve mollusks have shells divided into two halves. There are about 15,000 species of bivalves. Clams, oysters, mussels, and scallops are some of the most commonly known types. Most bivalves live in the ocean, but a few species live in fresh water. They can be tiny or very large. Giant clams can weigh more than 500 pounds (227 kg)!

Bivalves have a nervous system, a digestive system, and a heart. They have gills instead of lungs. A hard shell protects the soft body. Muscles connect the two halves of the shell, allowing it to close quickly and tightly. The shell opens when these muscles relax.

Bivalve species behave in different ways. Oysters and ocean mussels attach themselves to solid surfaces. Scallops swim by clapping the halves of their shell together. Clams and freshwater mussels use their "foot" to dig into the sand.

Each ring on a scallop's shell represents a year of growth. It can also represent a stressful event in the scallop's life.

compareandcontrast

Compare and contrast the different ways each bivalve uses its body to survive. How do their shells help them in similar ways? How do they help them in unique, or one-of-a-kind, ways?

When an oyster feels threatened, it will shut its shell to protect its soft body.

Gooey Gastropods

Gastropods make up the largest group of mollusks. The group contains more than 65,000 species! They include snails, slugs, and sea slugs. Most gastropods have one shell. The shells are commonly coiled and can have beautiful colors and patterns. Slugs—both on land and in the sea—do not have a shell.

Most land snails are hermaphrodites, meaning they have both male and female reproduction parts.

sea angel

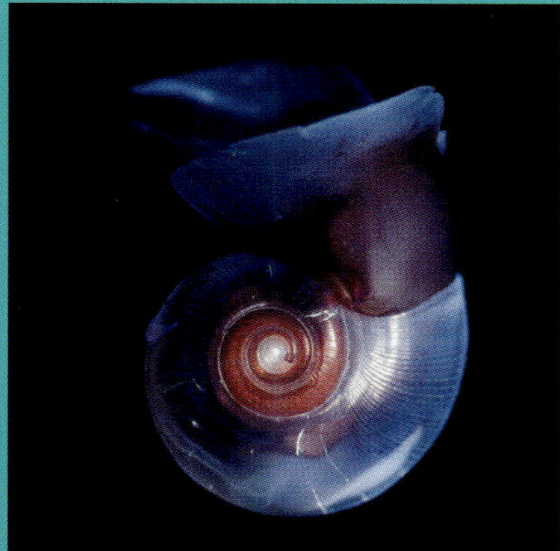

sea butterfly

Sea angels and sea butterflies are both gastropods. Sea angels do not have shells, but sea butterflies do.

Gastropods have a head and a soft body. They have a complete digestive system and highly developed sense organs. On land, snails and slugs release slime, which they glide over as they move.

Gastropods are found throughout the world. Some land gastropods spend the winter underground. Gastropods feed using a structure called the radula. The radula is covered with rows of tiny teeth. These tiny teeth help gastropods with grinding their food and scraping it off surfaces.

Consider This

What body parts and behaviors do snails have that help them adapt to living on land?

Underwater Cephalopods

Cephalopods are very unique-looking mollusks. There are about 650 species of cephalopods. The group includes octopuses, squid, nautiluses, and cuttlefish. These mollusks are active predators, feeding on fish and crustaceans.

The nautilus is the only cephalopod that has an external shell. Its shell is hard, coiled, and multi-chambered.

Some of a cephalopod's predators include sharks, seabirds, marine mammals, and other, larger cephalopods.

Cephalopods are very smart creatures. They have a big head, which contains a large brain. They also have a complex nervous system, or one having many parts, and highly developed eyes. Cephalopods commonly have eight to ten tentacles. The tentacles are covered with suckers, which help the cephalopod catch prey or attach to hard surfaces.

Generally, cephalopods do not have a shell. Instead, they protect themselves by swimming and moving quickly. Squid and octopuses can change color and patterns to blend into their surroundings. Most can also squirt an inky substance when an enemy is near. This makes the nearby water cloudy, allowing the cephalopod to distract, or confuse, a predator and escape.

Consider This

Most octopuses have the ability to camouflage, or change the color of their body to match their surroundings. How might this help them protect themselves from predators?

Traveling Homes

Mollusks live in many different places around the world. Most live in oceans, which are salty, but others live in fresh water. Some mollusks like shallow water along the coast. Others prefer deep water—some have been found at depths of 2,200 feet (670.5 m)! Many mollusks dig into sandy or muddy bottoms, but some attach themselves to rocks. Oysters, for example, grow in huge stacks called clusters. Most mollusks are free-living species, which means they do not have a permanent home. Some mollusk species are parasites and live on or inside another living thing.

Gastropods are helpful to your garden! Many species of slugs and snails eat decaying matter, which recycles the materials found in a garden.

Many people search marine and freshwater areas, such as coastlines, for mollusk shells. Many of these shells are highly valued by collectors.

Mollusks also live on land and even in our backyards! A garden is a perfect home for snails and slugs. These **terrestrial** mollusks generally like cool, moist places. But they can also be found in colder regions and in dry deserts.

WORD WISE

TERRESTRIAL REFERS TO ANYTHING THAT LIVES ON OR IN THE LAND, OR GROWING FROM THE LAND.

Ecosystem Helpers

An ecosystem is a community of interacting organisms that rely on each other to survive. Mollusks are an important part of their ecosystem. Terrestrial gastropods like snails and slugs put nutrients back into the soil as they eat. This keeps their habitat healthy.

Bivalves are also an important part of their ecosystem. They play a big part in cleaning Earth's waters by filtering sediments out of the water as they eat. One adult oyster can filter about 50 gallons (189 L) of water a day! This makes the water clean and clear—a healthy home for other marine creatures.

Cephalopods are important to their ecosystem because they are predators. They feed on fish, crabs, lobsters, and other marine animals, which keeps the ecosystem balanced. A balanced ecosystem helps ensure the survival of each member.

Mollusks that live on land play an important part in keeping soil healthy.

compare and contrast

A balanced ecosystem means every living thing is playing a role in keeping the environment healthy. What might an unbalanced ecosystem look like? How might it impact every member?

Cephalopods grow fast, and they don't live very long. This does help them adapt quickly to changes in their environment.

From Birth to Death

The first stage of life for bivalves is a fertilized egg. In most marine bivalves, the egg hatches into a larva. Larvae are wormlike creatures that look very different from the adult animal. Most bivalves have two larval stages. During these stages, the larvae are free swimming. After some time, the larvae fall to the sea floor and attach to a hard surface. When they become adults, they no longer swim.

Gastropods begin life in a similar way to bivalves. The fertilized egg hatches into a free-swimming larva. During the larval stage, the gastropod's body twists to bring the lower half of the body to the top. This is called torsion and only occurs in gastropods. They then grow into adults.

Female cephalopods lay clusters of eggs on the ocean floor, under rocks, or on hard surfaces. Some females protect the eggs until they hatch. After hatching, the young may resemble small versions of their parents, or they may enter a larval stage.

This mother octopus protects her eggs. Female octopuses usually die around the time their eggs hatch because they spend their time caring for the eggs rather than hunting for food.

compare and contrast

How are the life cycles of these three major mollusk classes similar? How are they different?

A female razor clam can lay half a million to 2 million eggs per spawn, or the process of laying eggs.

Mollusk Meals

What do mollusks like to eat? Many mollusks eat mostly algae. Gastropods use their radula to scrape algae off rocks. Larger mollusks have a bigger appetite. Snails and slugs eat plants and fruit. Squid may eat fish and shrimp. An octopus can go after prey as large as a shark!

Snails and slugs hide during the day in cool, moist areas such as under leaves, stones, or grass. They do this to avoid bright sunlight and heat.

Cuttlefish eat other mollusks such as octopuses and other cuttlefish.

Bivalves are filter feeders. That means they eat by straining tiny particles of food from the water. When bivalves open their shell, water washes over the gills. The gills then strain out oxygen and food.

Mollusks usually are active at night. Octopuses stay in the deep ocean during the day. At dawn and dusk, they swim closer to the surface looking for food. Giant squid use their foot-wide eyes to see prey in the dark ocean. Terrestrial snails and slugs are nocturnal, meaning they hide and rest during the day and come out at night to eat.

Consider This

How does a mollusk's body help it to hide, hunt, and feed?

Coexisting

Mollusks and humans have coexisted for a long time, and they help us in many ways. Many mollusks are a source of food for people around the world. Clams, oysters, scallops, octopuses, and squid are the most-commonly eaten mollusks. In France, escargots—or snails—are a delicacy.

Have you ever eaten clam chowder? If so, then you've eaten a mollusk!

Some oysters produce pearls within their shells. The odds of finding a pearl in an oyster are very rare . It's estimated to be a 1 in 10,000 chance!

Mollusks are important to the economy of many places along seacoasts. Mother-of-pearl, a shiny material on bivalve shells, is used for buttons and decorations. People use mollusk shells to make jewelry and other objects. Certain species of oysters make beautiful pearls, a valuable gemstone. If you've ever collected seashells on the beach, then you've probably held part of a mollusk!

Consider This

Humans have used mollusks as food and as part of the economy for years. If the mollusk population declined, how would humans be affected?

Threats to Survival

Earth has been home to mollusks for millions of years. In fact, some mollusk species are among the oldest-known animals. Mollusks have survived because their bodies are very well adapted to their environment. However, Earth's climate is changing, and the changes are affecting mollusks.

Global warming is causing the air temperatures on Earth to rise. As the planet warms, water temperatures also rise. Mollusks that thrive in cool temperatures are struggling to survive in warmer waters because their bodies have not adapted.

Scientists believe many species of marine mollusks will go extinct, or die out completely, because of global warming.

Declines in bivalve mollusk species have been blamed on overfishing. Overfishing refers to catching bivalve mollusks (or other creatures) at a rate faster than they can reproduce.

Human behavior is responsible for other problems that are negatively, or badly, affecting mollusk habitats too. For example, overfishing and pollution are causing the decline of mollusk populations.

WORD WISE

GLOBAL WARMING IS THE LONG-TERM HEATING OF EARTH'S SURFACE, CAUSING AN INCREASE IN TEMPERATURE. WHEN HUMANS BURN COAL, OIL, AND OTHER FOSSIL FUELS, GASES ARE RELEASED INTO THE ATMOSPHERE AND TRAP ENERGY FROM THE SUN. THIS CAUSES THE ATMOSPHERE TO WARM.

Protecting Mollusks

Our actions can have a negative impact on mollusks, and it's important that we are aware of ways to change this. Mollusks are a major source of food for many people and have played a large role in the economy of seaside places. Mollusks also help keep Earth's ecosystems balanced. We depend on mollusks in many ways, so we also need to be willing to protect them.

A person's carbon footprint refers to the amount of fossil fuels they emit into the atmosphere through their actions. Even everyday tasks like driving to school or work can increase a person's footprint, so it's important to find ways to use reusable energy, such as solar power.

These people are helping to clean the coast after an oil spill, which will aid in the survival of living creatures in the area.

It's never too early to begin learning about how to make Earth a better place for mollusks. You can start by thinking about the impact you have on the environment. Talk to your teacher and family about ways to keep our planet safe, clean, and balanced for mollusks and other living things, as well as ways to lessen your carbon footprint. A safer world for mollusks helps humans too.

Consider This

What are some ways you can lessen your carbon footprint? How will this help mollusks?

Glossary

adapt To change to fit a new situation.

characteristic A quality or appearance that makes an individual or group different from others.

climate The average weather conditions of a particular place or region over a period of years.

cluster A number of similar things growing, collected, or grouped together.

digestive system The parts of the body that work together to change food into a form that the body can use.

diverse Differing from one another.

muscle A body tissue consisting of long cells that can contract to produce motion.

nervous system The organ system that carries messages from the brain to other parts of the body.

nutrient Tiny substances in food that living things eat, drink, or absorb to grow and survive.

organ A part of the body that consists of cells and tissues and is specialized to do a particular task.

pollution The action or process of making land, water, or air dirty and not safe or suitable to use.

prey An animal that is hunted or killed by another animal for food.

varied Having many forms or types.

For More Information

Books

Dufresne, Emilie. *Endangered Animals in the Sea*. New York, NY: PowerKids Press, 2022.

Gaertner, Meg. *Snails*. Minneapolis, MN: Cody Koala, an imprint of Pop!, 2019.

Tyler, Madeline. *Mollusks*. New York, NY: KidHaven Publishing, 2020.

Websites

Mollusks for Kids
www.youtube.com/watch?v=r9gO-I1FVIQ
Watch this video for a fun review about the three main classes of mollusks.

Monterey Bay Aquarium: Cephalopods
www.montereybayaquarium.org/animals/animals-a-to-z/cephalopods
Read more about the underwater class of mollusks known as cephalopods and look at awesome pictures of them.

National Geographic Kids: What Is Climate Change?
www.natgeokids.com/uk/discover/geography/general-geography/what-is-climate-change
Learn more about the causes and impacts of climate change, as well as what you can do to make a difference.

Publisher's note to educators and parents: Our editors have carefully reviewed these websites to ensure that they are suitable for students. Many websites change frequently, however, and we cannot guarantee that a site's future contents will continue to meet our high standards of quality and educational value. Be advised that students should be closely supervised whenever they access the internet.

Index